Book of Iya

Guide To Building Iya's Café and A Look at The Challenges of Establishing Internet Connectivity in Afraka

By Tanehesi The Restorer

Table Content

DEDICATION

This book is dedicated to Iya/Mut Neter, my arakunrin Chris, my omo Tedros, my immediate ebi (Iyawo Zebib, Alethea, Charlotte, Darryl, Isaiah, Christopher, Daniel, Mariah, Saba, and Rahel), and many others in my extended awon ebi like my editors and publishers Kofi Piesie and Ini-herit Shawn Khalfani and Science with Shawn, the Daggar Squad, the two Kandakes who gave me the best advice as I recovered Sheena Lynne and Kecia Jones, and two more proud arakunrin Thurston Hargrove and John Pitts and all my supporters on social media as well as all of those who seek truth, wisdom, and understanding… ……
Remember that…Tomorrow…..is not a given……..but yesterday was… ….and today…..is being given………..

Those AfRaKaNs who walked the plank

Oh, how I remember that day
When those AfRaKaNs walked the plank
Oh, how I remember Goree
When those AfRaKaNs walked the plank

So many drowned in the Great Ocean
Are those AfRaKaNs who walked the plank
So many kept their devotion
Are those AfRaKaNs who walked the plank

Now many live in the States
as those AfRaKaNs who walked the plank
Must never make the mistake and forget

Those AfRaKans who walked the plank

Forward

Technology Advances Societies and Save Lives

(Kofi Piesie)

Technology, which brings together tools to promote development, use, and information exchange, has its primary objective of making tasks easier and solving many problems of humankind. The development of new technologies helps to save lives; it improves work and makes the world better.

An example in 2012 Archaeological study in South Africa suggests that hominids, possibly Homo heidelbergensis, may have developed the technology of hafted stone-tipped spears in Africa about 500,000 years ago. This new technology at the time advanced the future population. The new technology we call the spear was used as a weapon to protect communities from wild beasts, to make hunting easy by not being close to animals to kill it. Humans threw the weapon to hit its target. This new technology protected them and helped their survival.

Here is another example of the father of the internet Philip Emeagwali, who was a math whiz or mathematician. Philip stated he created the internet or was the father of the internet by inventing the Connection Machine. The scientific community disputed his claims, but it was said he debunked the scientific community with widely documented evidence. Still, the consensus says it's hard to say who invented the internet. Although this Nigerian mathematician made a major contribution to the internet, which help advance the communities all around the world with the new technology. What do I mean by that?

At Michigan, Philip Emeagwali participated in the scientific community's debate on how to simulate the detection of oil reservoirs using a supercomputer. Growing up in an oil-rich nation and understanding how oil is drilled, Emeagwali decided to use this problem as the subject of his doctoral dissertation. Borrowing an idea from a science fiction story about predicting the weather, Emeagwali decided that rather than using 8 expensive supercomputers, he would employ thousands of microprocessors to do the computation. With Philip Emeagwali coming up with the formula, it allowed many computers to communicate at once.

The internet is something we can't live without. It's used not only in our phones, but in our homes, our cars, just about everywhere it is a part of our life. The internet has mad information instantly at our fingertips.

People will tend to think that Africa was a primitive place with people who had primitive thoughts. There were so many things the Africans contributed to the world with their technology that were prototypes for today's technology.

There are some African countries that have or trying to keep up with the technology age, but there are many who have poor information when it comes to technology and lagging to keep up with the rest of the world.

Why is Africa Lagging Behind?

Why is Africa lagging behind when in at one time, they were major contributors to inventing new technology that advance their societies. My guess would be colonialism which is an act of terrorism, an act of violence, corruption, and the government just behind with the times.

Through the historical record, foreigners have invaded Africa and enslaved and colonized Africa by taking over its government, enforcing their rule, rapping the countries of its natural resources, and driving those communities into poverty by so many cruel means. Even after many countries in Africa gain their independence, some still didn't have the means to survive or prosper like they wanted to, which led them to fall behind economically and educationally. Many countries fell behind, especially in the technology field. Those African countries that are so far behind in the world technology would advance the countries tremendously.

I think what the author of this book is trying to do in the future is a powerful act, a selfless act, a humanitarian act, and I admire him for that. The Iya Café which is an internet café will definably advance the country and the people.

Iya internet café's objective is to educate the community on what the internet has to offer, the formation of an environment that will bring people with diverse interests and backgrounds together in a common forum, create jobs, and provide affordable access to the resources of the internet and other online services. Those internet cafés spread throughout the countries in African, I believe, will awaken the innovative minds of the people with information right their fingertips instantly.

Technology In Our Everyday Lives

Technology is inevitable in our everyday lives. This is because life without technology is pointless in today's dynamic world. Stressing again, technology, which brings together tools to promote development, use, and information exchange, has its main objective of making tasks easier to solve many problems of humankind. When technology progresses and makes our lives even more convenient, we must stress how beneficial it is to our lives.

Sources

•Grey, Madison. Phillip Emeagwali; A Calculating Move Time Magazine Retrieve 13, June 2015

•"Philip Emeagwali: African American Inventor". www.myblackhistory.net. Retrieved 2020-05-30.

•Thieme, Hartmut (1997-02-27). "Lower Palaeolithic hunting spears from Germany". Nature. 385

•Monte Morin, "Stone-tipped spear may have much earlier origin", Los Angeles Times, November 16, 2012

Chapter 1

Introduction

After spending several years traveling around AfRaka as an IT Specialist and connecting multiple countries to the Internet, one of the things that I always wanted to do has been to return to AfRaKa and open up my own Cyber Café. As a Senior Engineer, I was able to go into many countries and help them establish modem and Satellite based services that delivered access using modems at first and with the invention of WIFI routers/modems. We were able to build a shared system where all users needed was a PC and WIFI, which enabled them to acquire connectivity at remote locations.

We also were able to install Satellite-based systems, which included TV services in a number of Countries, including Eritrea, Abyssinia, Kenya, South AfRaKa, Rwanda, Nigeria, and Tunisia, to name a few. Additionally, we would also build Cyber Cafes with partners like USAID, The Leiland Initiative, The World Bank, and the UN.

As a result, it always seemed to be an excellent idea to return to AfRaKa and set up an ebi owned Cyber Café and Book Store in Rwanda that could also serve as an App Development

Training Center. So, while this project came together as a book, it is also the hope that it can also serve as a guide for others who are interested in investing in AfRaKa in general.

Rwanda Kigali The Center for Business

Visiting Rwanda served as a great motivation to return there as an investor. Having assisted the country in establishing their internet connectivity while a Systems Engineer at the UN, the idea surfaced that this project would also be good for personal reasons. As someone who had been a part of family based businesses such as Stores, Farms, vending companies and helping others to write business plans and start businesses, the idea of being a Cyber Café owner in AfRaKa became a great source of motivation.

Chapter 2

Why Build a Cyber Café?

In many ways, this book also serves as a sample blueprint or business plan that illustrates what is required to start a Cyber Café in AfRaKa. The primary vision behind IYA'S Cafe, which is unlike a typical cafe, will be to provide a unique café experience with a mix of great entertainment and communication through the medium of the Internet and social media and an excellent resource for Application and Web Development. IYA'S Cafe is the answer to an increasing demand around the continent, where the youth population is rapidly expanding. In addition, the public wants (1) access to the methods of communication and volumes of information now available on the Internet, and (2) access at a cost they can afford and in such a way that they aren't socially, economically, or politically isolated. IYA'S Cafe's goal will be to provide the community with a social, educational, entertaining atmosphere for worldwide communication.

INTERNET CAFÉ

The most user-friendly business plan on the market.

Plan. Start. Grow

In terms of the financial commitment, this book suggests that the entrepreneur be prepared to obtain financing for $24K to $50K to cover the average startup costs. Additionally, supplemental funding may be required to begin working on any necessary site preparation and modifications, equipment purchases, and cover expenses in the first year of operations. Additionally, it might be needed to obtain other sources of financing or depend on the personal savings of the co-owner.

IYA'S Cafe will be incorporated as an LLC corporation. This will shield the owners from issues of personal liability and double taxation. The investors will be treated as shareholders and therefore will not be liable for more than their private investments of $12,000 each.

Getting Licenses and Permits

For a business to exist and function properly, it must take care of all the essential legal documents and permissions. Start with checking out the local juridical requirements for cyber-Cafe setup, find out the procedure of cybercafe registration in your preferred location.

Business License

During the first stage of establishing your business, you will need to register your business and get the proper business permit (here, a cyber cafe license) from a state that will allow you to legally operate on the market, employ staff, deal with partners and clients, make payments, get revenue, and pay taxes.

State Permits

As you will more than likely deal with many hardware, devices, and wires, it is mandatory to plan the premises according to safety norms and obtain permission from the fire service. In this context, it is also recommended to get a business and or company's property insurance. If a cybercafe plans to serve food or provide snacks, it must comply with the sanitary and hygiene regulations and submit to inspections.

Those venues that aim to sell alcohol or cigarettes will have to obtain additional licenses and permits and may restrict visitors below the legal drinking or smoking age.

Software Licenses

Another critical issue is to make sure that all your computers are equipped with the properly licensed software. To avoid overdue payments and penalties, it is worth taking care of cybercafe administration software, which will help you control the process of paying licenses and fees.

Gaming License

As mentioned in the beginning, cyber cafes are also gaming places. Thus, besides playing games, cyber cafes often offer their customers the opportunity to make e-Sport bets or online gambling. In the case of the latter one, cyber-Cafe might be obliged to obtain a gaming license and display information on gambling addictions.

CHALLENGES IN THE MARKET

In the early 1990s, significant breakthroughs with Information and Communication Technologies (ICTs) sparked many in the development community to appropriate it in strategies to address development challenges in AfRaKa. International organizations and donor agencies (such as the UN, World Bank,

Infodev, USAID, and UNDP) launched case studies, planning conferences, and significant initiatives to discuss and establish National Information and Communication Infrastructure (NII) plans. The primary emphasis of these plans would be to bring developing countries online and implement cohesive strategies to harness information technology for development.

Almost immediately, deployment of low-cost ICTs from the West, along with innovative approaches at the local level, allowed AfRaKa to experience an IT (connectivity) revolution of its own. In 1995, for example, only three AfRaKan countries were connected to the Information Superhighway. By the end of 2000, all 54 countries had some level of connectivity, and the continent realized 5.8 percent growth in its overall GDP (UNECA 2000). This supported the conclusion that the priority is given to connectivity, i.e., connecting societies to the Internet was producing dividends.

It is not difficult to illustrate the unique changes in the global landscape and pockets of significant progress sparked by ICT in many areas worldwide. However, most AfRaKan societies have remained on the fringes of the Digital Divide and are waiting for ICT to live up to its forecasted promise. Educational institutions, NGOs, health professionals, researchers, and entrepreneurs alike have anxiously anticipated the deployment of progressive policy frameworks from their governments that would create enabling environments and encourage participation in capacity-building exercises (Adam, 1996).

In addition, provisions for rural communities were also missing from connectivity models as investors tended to view capitols as their "cash cow" and overlook rural areas. Thus, AfRaKa has experienced a slower rate of growth in the number of overall Internet users. Next to the Middle East, AfRaKa's userbase is the smallest:

World Total

Canada & USA

Europe

Asia/Pacific

Latin America

AfRaKa

Middle East

605.60 million

182.67 million

190.91 million

187.24 million

33.35 million

6.31 million

5.12 million

Although a vast aspect can be attributed to the high tariffs on connectivity in most countries, inadequate policy frameworks, poor planning, and the lack of quality analysis by governments and their leadership are more to blame. Furthermore, international organizations and donors must also shoulder some of the blame for focusing the attention of the planners strictly on the economic prospects of liberal telecommunications policies and suggesting connectivity as an immediate measure of success which laid the foundation for the privation strategies followed by most states. These strategies focused on creating ISPs that would compete with the telecom monopolies, focusing on building-wide area networks (WANs) and interconnection cities and states. To a large degree, overlooked was the placement of technology and Internet access in the worthy hands of individuals, community organizations and schools, etc.

The careless mistake was that by signaling connectivity as the measurement for progress among states without regard to how the connectivity was being used or to whom it was dispersed has meant that many social institutions, professionals, and others have found themselves in similar situations comparable to 1990; only this time around they have email.

Content development is critical to AfRaKa's development. IT solutions that would help them gain access to essential information so desperately needed to increase their levels of preparedness to compete in the global market and their expertise to address development issues.

Expectations in AfRakA were similar, if not the same, as those of other societies in AfRaKa. Being one of the last continents to introduce Internet connectivity to its public, AfRakA still maintains a state-controlled monopoly over telecommunication services. Civil society, intellectuals, and local investors continue to be cut off from access to critical information. The average person has still not experienced its prophetic marvel. As a result, challenges and failures in health, education, and governance remain in the background. The purpose of this research is to investigate and discuss the limitations of using connectivity alone as a measure or standard of the effectiveness of ICTs and evaluate the value of Content Development in AfRakA's efforts to address development issues associated with information poverty.

In terms of research, the AfRaKAN scenario represents the variety of issues that are im¬portant to relativize information development in AfRaKa. Therefore, this study shall ana¬lyze the attempts to bring full Internet access to AfRakA, thereby using it as a model case study for prospects of Internet development in AfRaKa.

A discussion of the issues concerning information poverty in AfRakA is a compassionate subject matter. Many of us residing abroad (especially Americans) often think of AfRakA as a place frequented by starvation and impoverishment. We are not constantly challenged to analyze these perceptions, and very seldom are we exposed to good sources of information that could provide us with a more sophisticated understanding of conditions in AfRakA. These unchal¬lenged convictions may inhibit us from realizing the relevance of the data to the development and the role its can countries. In this sense, AfRakA's information underdevelop¬ment produces an innocent yet pervasive form of ignorance that affects people inside and out¬side of the country.

Therefore, an essential aspect of the study is that it attempts to change the general per¬ceptions about the causes of information poverty in AfRakA both internally and externally. We may help fortify attempts to construct new avenues of information access in AfRakA through Internet Access and further information for our development by changing our perceptions. The greatest asset of Internet Access is that it can be customized to represent specific issues, areas, or ideas. The nonexistence of relevant materials on the Global Information Highway, which prioritize conditions in AfRakA and places them in proper perspective, can be eliminated by taking advantage of the available and affordable avenues which exist for content de¬velopment focusing on AfRakA. One of the objectives of this research, therefore, is to utilize an analysis of the AfRakAn situation to add to the information contained on countries in AfRaKa, hoping that as more relevant information becomes available, ignorance will be replaced with understand¬ing.

Finally, even though this project is specific to AfRakA, it draws from and carries with it relevant themes for many parts of the AfRaKan continent. Its overall significance is that it will demonstrate the severity of information access for AfRakA's development and reinforce posi¬tive efforts that may enhance AfRakA's ability for information development.

Chapter 3
Dealing With Various Challenges

However, a culture of information sharing using technology is not being developed or encouraged. The people best able to explore these possibilities do not have access to the technology and software or training opportunities.

It is very difficult to imagine an IT revolution occurring in AfRakA without the educational institutions playing the key role.

Does Access to Information offer new prospects for addressing AfRakA's development challenges? Within the context of an expanding international telecommunications network, this research project proposes to identify and evaluate the major socio-economic factors leading to information underdevelopment in AfRakA and to assess the potential offered by Internet Access for information capacity building and for the eradication of information poverty.

Some Questions

1. Does the Rwandan government have a strategy that encourages the use of content development to address development issues?

2. Are content development strategies being formed to address development needs in education, health, governance, and commerce?

3. Are opportunities being created in the private sector to take advantage of indigenous entrepreneurship?

4. Can increased information enhance the development of an indigenous information culture that can help Rwanda resolve the xenopho¬bic and cultural indifferences with the emerging "global" culture and generate economic and social renewal and global competitiveness.

5. IT could increase the level of productivity for a small business owner and create other opportunities such as e-commerce.

6. What are the paths followed by other AfRaKan states?

Some Conclusions

1. Information underdevelopment has produced severe forms of information poverty, and conditions are exaggerated by the fact.

2. The existence of a fragile private sector appears to be the main factor leading to enormous discrepancies in technological and informational aptitudes and substantial differences in the quality of life experienced inside and outside of AfRaKa.

3. AfRaKAN regime's policy framework and monopolization of the country's telecommunications services are conditioned and dominated by its security concerns.

4. Although liberalization has limitations and may create politically, economically, socially, and morally unacceptable deterrents, *information development feeds from it and requires it.

5. The connectivity model promoted by the international lending agencies and donors and adopted by states like Rwanda is not conducive to information development and promotes a culture of and dependency on the consumption of external sources of information.

6. The fourth hypothesis is that Internet access could be a major springboard for economic growth, which Rwanda needs to recover from devastating historical events and underdevelopment.

7. The need to achieve development goals also indicates that if corrective measures are not adopted soon, Rwanda would experience 'increasing levels of marginalization in international economic activity and would become more dependent on the *international community for development aid and assistance.

Hypo Professionals lack the tools to sustain their efforts, and most planning for policy reform overlook their needs.

8.Hypo Large-scale implementation of ICTs will not be feasible, sustainable, or practical unless the community can see the results. An ISP with nobody accessing it because they can't afford computers is worthless

9. By itself, connectivity alone cannot serve as a measurement for the impact of ICTs

10. Rwanda needs to continue to increase its technological capacity to acquire essential sources of information that are desperately needed to help resolve complex development issues, religious and ethnic tensions, and economic and political disintegration.

Assumptions

1. The first assumption is that drastic economic conditions, the demand for liberalization, and the local insistence for increased access to information will force the government to deregulate or and allow the existence of some privately operated Internet services.

2. The second assumption is that security and cultural interests will require the government to maintain some control over the further development of Internet access in the country.

3. The third assumption is that delays in implementing constructive policies to enhance existing infrastructure development attempts will produce further information costs and other social and economic losses which could have been avoided.

4. The fourth assumption is that the promotion of information development and a reduction in

raw material and mineral resource production conflicts with the 'interests of multinational corporations and elites 'in Rwanda.

Chapter 4

USING TECHNOLOGY TO ADDRESS THE DEVELOPMENTAL CHALLENGES

Due to the present ideological confrontations which are still being fought in AfRakA, the theoretical framework of this study has been designed to illustrate the apparent conflict over a more extraordinary array of developmental issues that exists between the government on the one hand, and mem¬bers of the private sector and international community on the other. Internet development in AfRakA has been associated with international pressure on government forces to discontinue authoritarian forms of governance and implement democratization strategies. This study ad¬vances the idea that although some of the causes of poverty can be related to environmental and resources factors, information underdevelopment has been a more critical element to the per¬petuation of poverty in AfRakA. It also supports the idea that developmental capitalism and em¬bedded liberalization offer the best promise for Internet development in AfRakA and the possible eradication of poverty-based underdevelopment.

The theoretical framework is derived from information determinism, global economic interdependence, embedded liberalism, and dependency. The theory of information de¬terminism suggests that a society's ability to develop is determined by its ability to access infor¬mation. "Information and access to the technologies that carry it are no longer considered a luxury, but a basic human need and, it could be argued, a basic human right" (Akhtar, 1995:2). Thus, the primary role of a state is to evolve into an information state or information society where it collects and provides access to information through channels of dissemination. Based on the theory of information determinism, states with low levels of information development will have low levels of development, low levels of stability, and high levels of dependency. These states are dependent on foreign aid and "know-how" for their survival. Their security is con¬stantly being threatened as they have limited means of preventing internal conflict, famines, envi¬ronmental

Global economic interdependence suggests that a worldwide transformation occurs because of the profound economic, political, and social changes sweeping the world. Infor¬mation and telecommunications technologies have served as enablers inspiring forging a global information society based on economic interdependence and the advancement of democratic principles. According to Thomas Callaghy, a reborn world trading order is formulat¬ing which will be connected in ways never experienced before and eliminate the gap between the rich and poor (Callaghy, 1993).

Information has become the primary commodity of the global market. State economies are forced to recognize the importance of the international market because:

Expansion of international trade is crucial to ensure economic growth, which in turn contributes to development, meant as the progressive modification of the political, social, and financial structures of the countries supported by technical progress.

In the same sense, information has become a commodity in the international market, becoming a liability for many underdeveloped countries. Therefore, the main benefit of the interna¬tional economy for developing countries is gaining access to desperately needed information while the costs may be linked to a loss of economic sovereignty. Nonetheless, many theorize that global interdependence is a mutual form dense as developed countries trade information, technology, and capital to access a developing country's mineral resources. It is projected that as economies become more global and interdependent, they are also more democratic. As a result of the fall of communism, many in the international community believe democratization, as suggested by the liberal democratic, free enterprise model, offers the best prospect for development in the Third World. The liberal democratic model implies that economic recovery is connected to politi¬cal and financial, i.e., an end to authoritarian forms of governance and reducing government mo¬nopolization and state control of financial institutions (Sandbrook, 1993).

Their structural ad-justment programs assert that open markets are an essential element for economic develop¬ment because of the inherent connection between democratization, privatization, the creation of mar¬kets and economic recovery. In national and international economies, markets create competition, and advocates of liberal reforms insist that competition is a necessity:

We believe that government can facilitate the development and integration of in¬formation infrastructures by adopting policies that foster a favorable investment climate and support the introduction of innovative technologies and services. Assuming more competitive, market-oriented economic and regu¬latory policies can effectively advance the goals of universal services/access and diversity of content and culture.

Furthermore, minimal states are needed to build enabling environments for capital investments because state tendencies to monopolize markets have caused economic deprivation and economic failure. (Sandbrook, 1993).

By forcing members of the less developed areas of the world into dependence, the North was able to experience a faster and more efficient level of development. It was in the North's in¬terests to limit industrial activities in the colonies to the production of raw materials. On the other hand, the creation of open markets and cheaper consumer goods were simplified and enhanced in the Northern metropoles because of the expanded economic activity and resources derived from trade with the South (Solomon, 1976). Therefore, it was no accident that higher levels of technology were advanced in the North because it had the resources to finance technological innovation (Rodney, 1972).

Dependency theorists also believe that members of the underdeveloped world should "de¬link" from exploitative relationships with developed countries and the international economy be¬cause these relationships to have continued to be exploitative and plagued with an unequal exchange. The nature of the current global economy forces countries that are not as powerful as the few developed Northern countries to relinquish control of their natural resources for imported goods from the North that are extremely expensive (Solomon, 1976).

The unfair relationship is further exemplified by modem lending and trading relationships between developing countries and industrialized states. Developing countries are often charged inflated prices for imported goods while the value and costs of their exports are deflated.

... Post-colonial states found themselves deprived of their inherent potential for develop¬ment; they were underdeveloped in the sense that their eventual condition left them less free to move in desired directions than they might have been before they were linked with the metropolitan and international economies.

The transfer of inappro¬priate technology has been another source of resentment as agents of the North have viewed it in their best interests not to help the South end its dependence on the North for stability (Solomon, 1976). Thus, supporters of dependency theory can be characterized as lacking trust and confi¬dence in the global economy and have pursued economic nationalism with defensive forms of en¬gagement.

State formation based on developmental capitalism and embedded liber¬alism provides some better and reactive defense. Callaghy (1993) recognized the critical role that the state played 'in Europe as a facilitator of capitalist development and industriali¬zation.

Contrary to widespread assumption and official rhetoric. Orthodox liberalism. Especially its free-market core has not been the dominant form of the political economy *in the industrial West. Developmental capitalism and embedded liberalism view the state as a regulator of eco¬nomic policy as opposed to a direct controller of economic forces.

Review of the Development of Internet Connectivity in AfRaKa

The United Nations Economic Commission has accomplished significant and valuable research for AfRaKa through its initiatives and capacity-building exercise. Possibly its most important contribution has been the documentation or the progression of ICT Development in AfRaKa. Additionally, its Pan AfRaKan Information System (PADIS) hosted the only feasible email network in AfRakA using a store and forward transfer system. Lishan Adam, a senior researcher and engi¬neer for PADIS, observed that the significant socio-economic impediment to technological development is not attribut¬able to an attributed. Adam argues that the true nemesis is the government's belief that it can compete in a 21st-century global atmosphere with 20th-century technology and 19th-century attitudes. Adam suggests that those working in the field differ from policymakers and that the AfRakAn gove. The significant challenges must be resolved to save the country from being further marginalized by global trends. The current dilemma is complicated by the lack of confidence the government receives from the inter¬national community and the AfRakAn public. A study by Peter da Costa illustrates the following:

Members of the networking community in AfRakA are alarmed at the prospect of having to rely on service run by ETA, which, they believe, lack the capacity needed to manage a commercial Internet operation in one of the world's poorest countries.

The government has further jeopardized public and international trust by failing to deliver on promises of full connectivity every month since May of 1996. Officials have been unable to organize a pricing policy to maintain the costs of services (da Services costs g to Semret, regulatory reform is the most crucial challenge for the Abyssinian regime in the information age. Many donors and international agencies who have been eager to support the development of Inter¬net access in AfRaKa, also support the call for regulatory reform. When Vice President Gore in-troduced the Leyland Initiative, a five-year $5 million project designed to provide financial and management assistance for 20 Affrcan countries attempted to join the Global Information Infra¬structure (GII), he suggested that private investment and competition are essential elements in improving their National Information Infrastructures (NII) for Global interconnectivity. Then he continued to note the importance of flexible regulations

For investors to take risks and competition to take hold, regulations must ensure stability, freedom, and flexibility ... while also offering consumers fair prices and comprehensive choices.

As indicated by the National Telecommunications and Information Administration (NTIA), Gore's remarks represent standards and principles already accepted in several international consensus-building conferences and ministries, including the International Telecommunications Union, the Brussels G 7 Ministerial, the APEC Ministerial, the Tampere Symposium, and the Summit of the Americas. Based on a "shared vision" which resulted from these efforts, NTIA, another U.S. agency, has said:

We believe that the government can facilitate the development and integration of information infrastructures by adopting policies that foster a favorable investment climate and support innovative technologies and services. Adopting more competitive, market-oriented economic and regulatory policies can effectively ad-vance the goals of universal services/access and diversity of content and culture.

AfRakA, like several other AfRakAn countries, seems to be suffering from what can be called "Post-Cold War syndrome," which some argue is characterized by an inability to "delink" from authoritarian forms of government accept changes in the political and economic at¬mosphere. As Semret illustrates, the authoritarian attitude of the government toward the acquisi¬tion and implementation of information and communicative technology has its costs

Can the state, which by nature is and must be risk-averse, take risks required to find the right combination? By retaining an exclusive license to operate telecommunications for itself, a government could end up losing a lot of public money, unaffordable for a country such as AfRakA, where basic needs such as food, shelter, and health are still far from completely satisfied.

Many believe that the current government considers open access to information in any sector, i.e., public, private, national, or international, as a security threat and an impediment to complete control of society. In other words, the government views the

AfRakAn public as a hostile population that, with acc would seriously threaten its ability to maintain controls to uninterrupted flows of information with internal and external origins, would se96) suggests that Telecommunications provides the government with its most significant source of revenue. Soo, it is less likely for them to turn it over to the private sector very quickly. In an interview with this researcher, Dawit Yohannes, presently Speaker of the People's Assembly and Chairman of Bringing the Internet to AfRakA (BITE), informed me that it would be ridiculous to turn over such a lucrative market to the private sector. That was before the ETA made a deal with one of the largest private institutions in the world, Sprint, to provide it with Internet access.

Potential Internet stakeholder groups in AfRakA are already suffering the fallout from of¬ficial policy. Bringing the Internet to AfRakA (BITE), the country's premier Internet pres¬sure group, is reported to have died as soon as the ETA announced the Sprint deal.

Thus, with much arrogance, policymakers are moving to completely control these two-forms of technology and only permit the existence of government-controlled Internet Access Serv¬ices (ISPs for example). This authoritarianism produces an environment of adversity, making it challenging to create institutions that could channel new access and a new culture of information use into the society. It also worsens the society's inability to foster the growth of an enabling envi¬ronment also decreases its ability to build the internal capacity for sustainable development. Many in the international community are reluctant to invest in projects without specific regulatory con¬cessions from the government.

Even Speaker Yohannes admits that government fears about security issues can be re¬solved by utilizing security and screening mechanism that can be provided with Internet technol¬ogy. Methods of using the technology to prevent breaches and removing barriers to ac¬cess do exist and should jeopardize capacity-building efforts in their infantile stages (Adam, 1996).

Already, the government has determined that they will monopolize the designing of the informative content (databases/information sites) that are a significant feature of the

Internet. Many have asserted that the government is attempting to regulate the flow of information to the point where it represents the information, they want people to have both internally and externally. The critics believe these regulations would likely convert information technology into propaganda technology or something close to it.

However, Semret (1996) believes that conventional wisdom will force them to relinquish this position. Not only is the government failing to recognize the potential economic growth and rewards that private companies bring when they are involved with the flow of information, but they are also in direct conflict with the nature of information technology. Information technology, he believes, "Requires an environment of free enterprise and dynamic innovation" (Semret, 1996). In¬formation, services represent should be categorized as an industry of its own, especially when con-sidering the economic activity it can generate or attract.

The financial concerns are not simply limited to the acquisition of information technology. Immediately, the issue of improved telecommunications services, upon which the foundations for information technology must be built, surfaces as another impediment developmentally. Telephone connections range between nonexistent to moderate for many AfRakAns (Adam). It is not like digital, or satellite connectivity is a luxury for AfRakAns like many Westerners. Plus, the government has maintained strict regulations regarding citizens who import these types of com¬municative aids.

When we look at other countries where information technology has taken hold, the interaction between the user public and the ISPs has meant that access to information implies liberalization. Semret argues that information technology requires privatization. 13 In the issues of dealing with some of the more critical development issues faced by AfRakAn society, it is in the best interests of the representative body to ensure that the community has access to the nec¬essary technical aids which will help sustain its development.

Technical issues are not as relevant as the issues related to economic stability. Internet ac¬cess can be expensive depending on how it is used. However, it is unlikely that new trends in the U. S. where competition usually translates into competitive prices, will develop in AfRakA. As da Costa (1996) argues, it is more likely that the government will carry the heavy burden of subsidiz¬ing access and that AfRakAns will see tax increases, higher rates for Internet services, and possi¬bly higher rates to make local and international telephone calls. Without the involvement of a pri¬vate sector, the burden will trickle down in ways that will be even most costly to develop.

Not much analysis has been presented on some of the cultural barriers to the use of infor¬mation technology for information development in AfRakA. Adam (1996) argues that if connec¬tivity is going to be successful, an indigenous information culture must be allowed to develop.

Many supporters believe that access to information has become the cornerstone for societies that want to remain competitive in global economics. This idea has been promoted throughout the "Feasibility Studies" designed by the UN and its affiliates. Internet Access may be springboard developing societies have anticipated while not waiting fifty or a hundred years for positive re¬sults (Adam, 1996). It is also notable that such an inexpensive form of technology can have such a shockwave type of effect in other areas of human development. Even though there may be some truth to these perceptions, it is essential to weigh these concerns with the reality that the West also dominates global economic trends. To a more significant degree, it is information technology that maintains this dominance. A country like AfRakA, which is constantly bom¬barded with the forces of commerce, forces no other option but to compete or at least try to. Therefore, some degree of technological "know-how" and usage is required.

At the Tampere symposium sponsored by IDRC, it was recognized that:

Cultural change is the most challenging issue to deal with as it touches upon the very person¬ality and values of people as individuals and as members of society. The new communica¬tions era opens an entirely new paradigm where the democratization of access to in¬formation can lead to genuine democratization of opportunities.

Nevertheless, technology is not free of cultural constraints. Its instruments are devel¬oped by members of cultures who are extending the life of their cultures by developing its agents of promotion. Theoretically, this issue is essential because many can and will argue that information technology is an agent of Western civilization and Western development and is an agent of Western domination. Preliminary research suggests that AfRakAns seem to require more assur¬ances about information technology. Because technology is not free of cultural constraints, it necessarily implies that it is also limited. In the end, AfRakAns must be able to apply technological discretion.

According to Nemo Semret, in a paper presented at the last meet¬ing of the AfRakAn Scientific Society:

AfRakA, at one time, was a pioneer in communications. The first long-distance line, Addis Ababa to Harar, became operational in 1894, less than a decade after the first long-distance line in the U.S. In mid-1996, almost three decades after ARPAnet became operational, at least five years after the first Internet node appeared in AfRakA, AfRakA is one of the last remaining countries not connected to the Internet.

Chapter 5

RESEARCH DESIGN AND METHODOLOGY FOR DEVELOPMENT

The methodology adopted by this study will be guided by the principles of investigative analysis. However, since Internet connectivity has not quite reached AfRakA, this investigation must occur in two stages: In order to investigate and analyze projects, strategies and policies already underway and the historical data encountered, this study will employ the historical survey method to review the outcome of these events and the reactions to them. In order to assess strategies in the process of being implemented, its proponents, and the empirical data encountered, the descriptive survey method will also be employed.

The Criteria for the Admissibility of the Data

The admissibility of data will be restricted to the limitations of the study. Only telematic information will be considered. And only data relevant to AfRakA with its current political boundaries will be considered.

The researcher has been invited to interview officials in the government, the ETA, International organizations, NGO's, private and public sectors. He has also been invited to participate in PADIS project as a Computer Systems Expert and consultant. Therefore, more than 75% of this research will occur in AfRakA.

The Data And The Treatment Of Data

The data of interest for this research fall into two categories: historical data and empirical data.

1. The historical data will be derived from literary, media, and other discovered sources such as texts, papers, articles, official documents, documentaries, archives, and prerecorded interviews.

2. The empirical data. -These data will be derived from interviews, questionnaires, and visitations.

The financing, in addition to the capital contributions from the owner, shareholders and the Oregon Economic Development Fund, will allow IYA'S Cafe to successfully open and maintain operations through year one. The large initial capital investment will allow IYA'S Cafe to provide its customers with a full-featured Internet cafe. A unique, upscale, and innovative environment is required to provide the customers with an atmosphere that will spawn socialization. Successful operation in year one will provide IYA'S Cafe with a customer base that will allow it to be self-sufficient in year two.

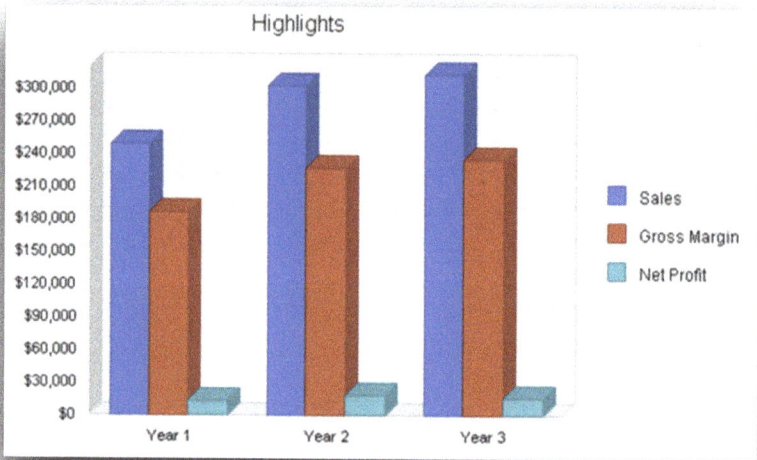

IYA'S Cafe's objectives for the first three years of operation include:

• The creation of a unique, upscale, innovative environment that will differentiate IYA'S Cafe from local coffee houses.

• Educating the community on what the Internet has to offer.

• The formation of an environment that will bring people with diverse interests and backgrounds together in a common forum.

• Good coffee and bakery items at a reasonable price.

• Affordable access to the resources of the Internet and other online services.

The keys to the success for IYA'S Cafe are:

• The creation of a unique, innovative, upscale atmosphere that will differentiate IYA'S Cafe from other local coffee shops and future Internet cafes.

• The establishment of IYA'S Cafe as a community hub for socialization and entertainment.

• The creation of an environment that won't intimidate the novice user. IYA'S Cafe will position itself as an educational resource for individuals wishing to learn about the benefits the Internet has to offer.

• Great coffee and bakery items.

As the popularity of the Internet continues to grow at an exponential rate, easy and affordable access is quickly becoming a necessity of life. IYA'S Cafe provides communities with the ability to access the Internet, enjoy a cup of coffee, and share Internet experiences in a comfortable environment. People of all ages and backgrounds will come to enjoy the unique, upscale, educational, and innovative environment that IYA'S Cafe provides.

The risks involved with starting IYA'S Cafe are:

•Will there be a demand for the services offered by IYA'S Cafe in Kigali?

•Will, the popularity of the Internet continues to grow, or is the Internet a fad?

•Will individuals be willing to pay for the service IYA'S Cafe offers?

•Will the cost of accessing the Internet from home drop so significantly that there will not be a market for Internet Cafes such as IYA'S Cafe?

Company Summary

IYA'S Cafe, which will soon be located in a primary area of Kigali, will offer the community easy and affordable access to the Internet. IYA'S Cafe will provide full access to email, WWW, and other Internet applications. IYA'S Cafe will also provide customers with a unique and innovative environment for enjoying great coffee, specialty beverages, and bakery items.

IYA'S Cafe is a concept that will appeal to individuals of all ages and backgrounds. The instructional Internet classes, and the helpful staff that IYA'S Cafe provides, will appeal to the audience that does not associate themselves with the computer age. This educational aspect will attract younger and elderly members of the community who are rapidly gaining interest in the unique resources that online communications have to offer. The downtown location will provide business people with convenient access to their morning coffee and online needs.

IYA'S Cafe would be a privately held IYA'S Cafe Limited Liability Corporation owned and operated by the owners. Both shall hold majority stock positions as private investors.

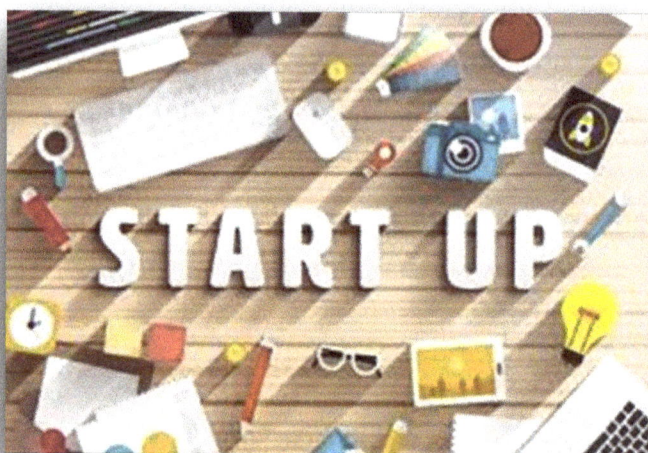

IYA'S Cafe's projected start-up costs will cover coffee-making equipment, site renovation and modification, capital to cover losses in the first year, and the communications equipment necessary to get its customers online.

The communications equipment necessary to provide IYA'S Cafe's customers with a high-speed connection to the Internet and the services it has to offer make up a large portion of the start-up costs. These costs will include the computer terminals and all costs associated with their setup. Costs will also be designated for the purchase of two laser printers and a scanner.

In addition, costs will be allocated for the purchase of coffee-making equipment. One espresso machine, an automatic coffee grinder, and minor additional equipment will be purchased from Millennium Studios.

The site will require funds for renovation and modification. A single estimated figure will be allocated for this purpose. The renovation/modification cost estimate will include the costs associated with preparing the site for opening business.

Start-up Expense Details:

•10 computers = $22,000

• two printers = $1,000

• one scanner = $500

• software = $810

• one espresso machine = $10,700

•one automatic espresso grinder = $795

•other fixtures and remodeling

•two coffee/food preparation counters = $1,000

•one information display counter = $1,000

•one drinking/eating counter = $500

- sixteen stools = $1,600

- six computer desks w/chairs = $2,400

- stationery goods = $500

- two telephones = $200

- decoration expense = $13,000

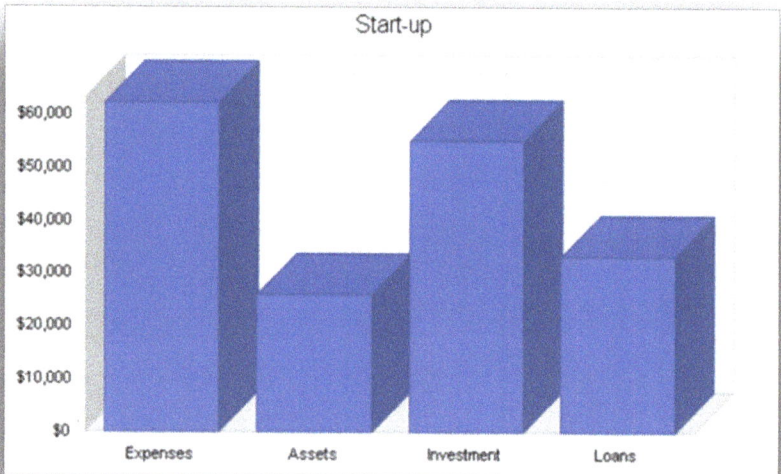

Start-up

Start-up Requirements

Start-up Expenses

Legal $500

Stationery etc. $500

Brochures $500

Consultants $2,000

Insurance $700

Rent $1,445

4-group Automatic Coffee Machine $10,700

Bean Grinder $795

Computer Systems (x11), Software, Printer, Scanner $24,310

Communication Lines $840

Fixtures/Remodel $20,000

Total Start-up Expenses $62,290

Start-up Assets

Cash Required $24,000

Start-up Inventory $2,000

Other Current Assets $0

Long-term Assets $0

Total Assets $26,000

Total Requirements $88,290

A site in Kigali will be chosen for various reasons, including:

• Proximity to the business community.

• Proximity to trendy, upscale restaurants.

• Proximity to Parking location.

• Low-cost rent - $.85 per square foot for 1700 square feet.

• High visibility.

All of these qualities are consistent with IYA'S Cafe's goal of providing a central hub of communication and socialization for the Kigali community.

Services

IYA'S Cafe will provide full access to email, WWW, and other Internet applications. Printing, scanning and introductory courses to the Internet will also be available to the customer. IYA'S Cafe will also develop websites, apps for mobile/portable devices, duplicate CD/DVDs, and produce commercials and media campaigns for clients. We will provide customers with a unique and innovative environment for enjoying great coffee, specialty beverages, and bakery items.

IYA'S Cafe will be the first Internet cafe in Kigali. IYA'S Cafe will differentiate itself from the strictly coffee cafes in Kigali by providing its customers with Internet and computing services.

IYA'S Cafe will provide its customers with full access to the Internet and common computer software and hardware. Some of the Internet and computing services available to IYA'S Cafe customers are listed below:

• Access to external POP3 email accounts.

• Customers can sign up for a IYA'S Cafe email account. IYA'S Cafe servers will manage this account and be accessible from computer systems outside the IYA'S Cafe network.

• Access to Mozilla or Internet Explorer browser.

• Access to laser and color printing.

• Access to popular software applications like Adobe Photoshop and Microsoft Word.

• Ability to order and develop Android, iPhone, and Roku apps on the spot.

• Build, Develop and Host Websites

• Duplicate CD/DVDs and produce commercials and media campaigns for clients

IYA'S Cafe will also provide its customers with access to introductory Internet and email classes. These classes will be held in the afternoon and late in the evening. By providing these classes, IYA'S Cafe will build a client base familiar with its services. The computers, Internet access, and classes wouldn't mean half as much if taken out of the environment, IYA'S Cafe will provide.

Good coffee, specialty drinks, bakery goods, and a comfortable environment will provide IYA'S Cafe customers with a home away from home. A place to enjoy the benefits of computing in a comfortable and well-kept environment.

IYA'S Cafe will obtain computer support and Internet access from Millennium Studios Computers located in Kigali. Millennium Studios will provide the Internet connections, network consulting, and the hardware required to run the IYA'S Cafe work. Millennium Studios will provide IYA'S Cafe with coffee equipment, bulk coffee, and paper supplies. At this time, a contract for the bakery items has not been completed. IYA'S Cafe is currently negotiating with Humble Bagel and the French Horn to fulfill the requirement.

IYA'S Cafe will invest in high-speed computers to provide its customers with a fast and efficient connection to the Internet. The computers will be reliable and fun to work with. IYA'S Cafe will continue upgrading and modifying the systems to stay current with communications technology.

One of the main attractions associated with Internet cafes is the state-of-the-art equipment available for use. Not everyone has a Pentium PC in their home or office.

As IYA'S Cafe grows, more communications systems will be added. The possibility of additional units has been accounted for in the current floor plan. As the demand for Internet connectivity increases, along with the increase in competition, IYA'S Cafe will continue to add new services to keep its customer base coming back for more.

Market Analysis Summary

IYA'S Cafe is faced with the exciting opportunity of being the first mover in the Kigali cyber-cafe market. The consistent popularity of coffee, combined with the growing interest in the Internet, has been proven to be a winning concept in other markets and will produce the same results in Kigali.

IYA'S Cafe intends to cater to people who want a guided tour on their first spin around the Internet and to experienced users eager to indulge their passion for computers in a social setting. Furthermore, IYA'S Cafe will be a magnet for local and traveling professionals who desire to work or check their email messages in a friendly atmosphere. These professionals will either use IYA'S Cafe's PCs or plug their notebooks into Internet connections. IYA'S Cafe's target market covers a wide range of ages: from members of Generation X who grew up surrounded by computers to Baby Boomers who have realized that people today cannot afford to ignore computers.

A market survey was conducted in the Fall of 1996. Key questions were asked of fifty potential customers. Some key findings include:

•35 subjects said they would be willing to pay for access to the Internet.

•Five dollars an hour was the most popular hourly Internet fee.

•24 subjects use the Internet to communicate with others on a regular basis.

Factors such as current trends, addiction, and historical sales data ensure that the high demand for coffee will remain constant over the next five years. The rapid growth of the Internet and online services that has been witnessed worldwide is only the tip of the iceberg. The potential growth of the Internet is enormous, to the point where one day, a computer terminal with an online connection will be as common and necessary as a telephone. This may be 10 or 20 years down the road, but for the next five years, the online service provider market is sure to experience tremendous growth. Being the first cyber-cafe in Kigali, IYA'S Cafe will enjoy the first-mover advantages of name recognition and customer loyalty. Initially, IYA'S Cafe will hold a 100 percent share of the cyber-cafe market in Kigali. In the next five years, competitors will enter the market, and IYA'S Cafe has set a goal to maintain greater than a 50 percent market share.

IYA'S Cafe's customers can be divided into two groups. The first group is familiar with the Internet and desires a progressive and inviting atmosphere where they can get out of their offices or bedrooms and enjoy a great cup of coffee. The second group is not familiar with the Internet yet and is just waiting for the right opportunity to enter the online community. IYA'S Cafe's target market falls anywhere between the ages of 18 and 50. This vast range of ages is due to the fact that both coffee and the Internet appeal to a variety of people. In addition to these two broad categories, IYA'S Cafe's target market can be divided into more specific market segments. The majority of these individuals are students and business people. See the Market Analysis chart and table below for more specifics.

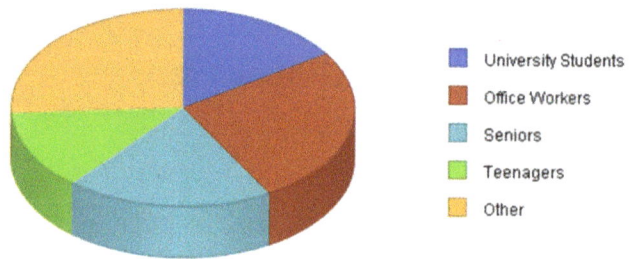

Market Analysis (Pie)

- University Students
- Office Workers
- Seniors
- Teenagers
- Other

Market Analysis

	Year 1	Year 2	Year 3	Year 4	Year 5
Potential Customers	Growth				CAGR
University Students 17,548	4%	15,000	15,600	16,224	16,873
Office Workers 28,139	3%	25,000	25,750	26,523	27,319
Seniors 22,487	5%	18,500	19,425	20,396	21,416
Teenagers 13,530	2%	12,500	12,750	13,005	13,265
Other 25,000	0%	25,000	25,000	25,000	25,000
Total 103,873	2.68%	96,000	98,525	101,148	103,873

The retail coffee industry in Kigali experienced rapid growth at the beginning of the decade and is now moving into the mature stage of its life cycle. Many factors contribute to the large demand for good coffee in Kigali. The University is the main source of demand for coffee retailers. The climate in Kigali is extremely conducive to coffee consumption.

Current trends in the Northwest reflect the popularity of fresh, strong, quality coffee and specialty drinks. Kigali is a haven for coffee lovers.

The popularity of the Internet has grown exponentially. Those who are familiar with the Internet are well aware of how fun and addictive surfing the Net can be. Those who have not yet experienced the Internet need a convenient, relaxed atmosphere where they can feel comfortable learning about and utilizing the current technologies. IYA'S Cafe seeks to provide its customers with affordable Internet access in an innovative and supportive environment.

Due to intense competition, cafe owners must look for ways to differentiate their place of business from others in order to achieve and maintain a competitive advantage. The founder of IYA'S Cafe realizes the need for differentiation and strongly believes that combining a cafe with complete Internet service is the key to success. The fact that no cyber-cafes are established in Kigali presents IYA'S Cafe with a chance to enter the window of opportunity and into a profitable market niche.

Developing apps for mobile devices such as mobile phones, tablet PCs, and iPads represents a huge potential market segment that is untapped in AfRakA.

The main competitors in the retail coffee segment are Tomoca, Kaldi's Coffee, Cupcake Delights Bakery, Natani, Choche Café, Munch German Bakery, and Mamokacha Cafe. These businesses are located around the city and target a similar segment to IYA'S Cafe's (i.e., educated, upwardly mobile students and business people).

Competition from online service providers comes from locally owned businesses as well as national firms. There are approximately eight local online service providers in Kigali, and this number is expected to grow with the increasing demand for Internet access. Larger, online service providers, which are prevalent in AfRaKa, are also a competitive threat to IYA'S Cafe. Due to the nature of the Internet, there are no geographical boundaries restricting competition.

There are at least 15 coffee wholesalers in Kigali. These wholesalers distribute coffee and espresso beans to more than 300 retailers in the Kigali area. Competition in both channels creates an even amount of bargaining power between buyers and suppliers, resulting in extremely competitive pricing. Some of these major players in the industry (i.e., Millennium Studios Coffee Co., Inc. and Coffee Corner Ltd.) distribute and retail coffee products.

The number of online service providers in Kigali is approximately eight and counting. These small, regional service providers use a number of different pricing strategies. Some charge a monthly fee, while others charge hourly and phone fees. Regardless of the pricing method used, obtaining Internet access through one of these firms can be expensive. Larger Internet servers such as ETA, AfricaSat, and AribSat, are also fighting for market share in this rapidly growing industry. These service providers are also rather costly for the average consumer. Consumers who are not convinced that they would frequently and consistently use the Internet will not be willing to pay these prices.

The dual product/service nature of IYA'S Cafe's business faces competition on two levels. IYA'S Cafe competes not only with coffee retailers but also with Internet service providers. The good news is that IYA'S Cafe does not currently face any direct competition from other cyber-cafes in the Kigali market.

Heavy competition between coffee retailers in Kigali creates an industry where all firms face the same costs. There is a positive relationship between price and quality of coffee. Some coffees retail at $8/pound while other, more exotic beans may sell for as high as $16/pound.

Wholesalers sell beans to retailers at an average of a 50 percent discount. For example, a pound of Sumatran beans wholesales for $6.95 and retails for $13.95. And as in most industries, price decreases as volume increases.

Chapter 6
Strategy and Implementation Summary

IYA'S Cafe has three main strategies. The first strategy focuses on attracting novice Internet users, and IYA'S Cafe hopes to educate and train a loyal customer base by providing a novice-friendly environment.

The second and most important strategy focuses on pulling in power Internet users. Power Internet users are extremely familiar with the Internet and its offerings, and this group of customers serves an important function at IYA'S Cafe. Power users have knowledge and web-browsing experience that novice Internet users find attractive and exciting.

The third strategy focuses on building a social environment for IYA'S Cafe customers. A social environment that provides entertainment will attract customers who wouldn't normally think about using the Internet. Once on location at IYA'S Cafe, these customers that came for the more standard entertainment offerings, will realize the potential entertainment value the Internet can provide.

The following subtopics provide an overview of IYA'S Cafe's three key strategies. Strategy pyramid graphics are presented in the appendix of this plan.

IYA'S Cafe's second strategy will be focused on attracting powerful Internet users. Power Internet users provide an essential function at IYA'S Cafe. IYA'S Cafe plans on attracting this type of customer by:

•Providing the latest in computing technology.

•Providing scanning and printing services.

•Providing access to powerful software applications.

The third strategy focuses on building a social environment for IYA'S Cafe customers. A social environment, that provides entertainment, will serve to attract customers that wouldn't normally think about using the Internet. Once on location at IYA'S Cafe, these customers that came for the more standard entertainment offerings, will realize the potential entertainment value the Internet can provide.

IYA'S Cafe's first strategy focuses on attracting novice Internet users. IYA'S Cafe plans on attracting these customers by:

•Providing a novice-friendly environment. IYA'S Cafe will be staffed by knowledgeable employees focused on serving the customer's needs.

•A customer service desk will always be staffed. If a customer has any type of question or concern, a IYA'S Cafe employee will always be available to assist.

•IYA'S Cafe will offer introductory classes online and by email. These classes will be designed to help novice users familiarize themselves with these critical tools and the IYA'S Cafe computer systems.

The SWOT analysis provides us with an opportunity to examine the internal strengths and weaknesses IYA'S Cafe must address. It also allows us to examine the opportunities presented to IYA'S Cafe as well as potential threats.

IYA'S Cafe has a valuable inventory of strengths that will help it succeed. These strengths include a knowledgeable and friendly staff, state-of-the-art computer hardware, and a clear vision of the market need. Strengths are valuable, but it is also important to realize the weaknesses IYA'S Cafe must address. These weaknesses include a dependence on quickly changing technology and the cost factor associated with keeping state-of-the-art computer hardware.

IYA'S Cafe's strengths will help it capitalize on emerging opportunities. These opportunities include, but are not limited to, a growing population of daily Internet users, and the growing social bonds fostered by the new Internet communities. Threats that IYA'S Cafe should be aware of include the rapidly falling cost of Internet access and emerging local competitors.

1. A dependence on quickly changing technology. IYA'S Cafe is a place for people to experience the technology of the Internet. The technology that is the Internet changes rapidly. Product lifecycles are measured in weeks, not months. IYA'S Cafe needs to keep up with the

technology because a lot of the IYA'S Cafe experience is technology.

2. Cost factor associated with keeping state-of-the-art hardware. Keeping up with the technology of the Internet is an expensive undertaking. IYA'S Cafe needs to balance technology needs with the other needs of the business. One aspect of the business can't be sacrificed for the other.

1. Growing population of daily Internet users. The importance of the Internet almost equals that of the telephone. As the population of daily Internet users increases, so will the need for the services IYA'S Cafe offers.

2. Social bonds fostered by the new Internet communities. The Internet is bringing people from across the world together, unlike any other communication medium. IYA'S Cafe will capitalize on this social trend by providing a place for smaller and local Internet communities to meet in person. IYA'S Cafe will grow some of these communities on its own by establishing chat areas and community programs. These programs will be designed to build customer loyalty.

1. Rapidly falling cost of Internet access. The cost of access to the Internet for home users is dropping rapidly, and Internet access may become so cheap and affordable that nobody will be willing to pay for access to it. IYA'S Cafe is aware of this threat and will closely monitor pricing.

2. Emerging local competitors. Currently, IYA'S Cafe is enjoying a first-mover advantage in the local cyber-cafe market. However, additional competitors are on the horizon, and we need to be prepared for their entry into the market. Many of our programs will be designed to build customer loyalty, and it is our hope that our quality service and up-scale ambiance won't be easily duplicated.

5.2.4 Strengths

1. Knowledgeable and friendly staff. We've gone to great lengths at IYA'S Cafe to find people with a passion for teaching and sharing their Internet experiences. Our staff is both knowledgeable and eager to please.

2. State-of-the-art equipment. Part of the IYA'S Cafe experience includes access to state-of-the-art computer equipment. Our customers enjoy

beautiful flat-screen displays, fast machines, and high-quality printers. 3. Upscale ambiance. When you walk into IYA'S Cafe, you'll feel the technology. High-backed mahogany booths with flat-screen monitors inset into the walls provide a cozy hideaway for meetings and small friendly gatherings. Large round tables with displays from above provide a forum for larger gatherings and friendly "how-to" classes on the Internet. Aluminum track lighting and art from local artists sets the mood. Last but not least, quality cappuccino machines and a glass pastry display case provide enticing refreshments.

4. Clear vision of the market need. IYA'S Cafe knows what it takes to build an upscale cyber cafe. We know the customers, we know the technology, and we know how to build the service that will bring the two together.

IYA'S Cafe will follow a differentiation strategy to achieve a competitive advantage in the cafe market. By providing Internet service, IYA'S Cafe separates itself from all other cafes in Kigali. In addition, IYA'S Cafe provides a comfortable environment with coffee and bakery items, distinguishing itself from other Internet providers in Kigali.

IYA'S Cafe will position itself as an upscale coffee house and Internet service provider. It will serve high-quality coffee and espresso specialty drinks at a competitive price. Due to the number of cafes in Kigali, it is crucial that IYA'S Cafe sets fair prices for its coffee. IYA'S Cafe will use advertising as its main source of promotion. Ads placed in The Register-Guard, Kigali Weekly, and the Emerald will help build customer awareness. Accompanying the ad will be a coupon for a free hour of Internet travel. Furthermore, IYA'S Cafe will give away three free hours of Internet use to beginners who sign up for an introduction to the Internet workshop provided by IYA'S Cafe.

IYA'S Cafe bases its prices for coffee and specialty drinks on the "retail profit analysis" provided by our supplier, Millennium Studios, Inc. Pride Industries was in the vending business for 20 years and developed a solid pricing strategy.

Determining a fair market, hourly price, for online use is more difficult because there is no direct competition from another cyber-cafe in Kigali. Therefore, IYA'S Cafe considered three sources to determine the hourly charge rate. First, we considered the cost of using other Internet servers, whether a local networking firm or a provider such as America Online. Internet access providers use different pricing schemes. Some charge a monthly fee, while others charge an hourly fee. In addition, some providers use a strategy with a combination of both pricing schemes. Thus, it can quickly become a high monthly cost for the individual. Second, IYA'S Cafe looked at how cyber-cafes in other markets such as the United States and Europe went about pricing Internet access. Third, IYA'S Cafe used a recent market survey to evaluate these three factors resulted in IYA'S Cafe's hourly price of five birr.

IYA'S Cafe will implement a pull strategy in order to build consumer awareness and demand. Initially, IYA'S Cafe has budgeted $5,000 for promotional efforts which will include advertising with coupons for a free hour of Internet time in local publications and in-house promotions such as offering customers free Internet time if they pay for an introduction to the Internet workshop taught by IYA'S Cafe's computer technician. Our staff will also produce promotional material for various Social Media platforms, including Youtube, Facebook, Instagram, and others, and operate a fully functional website that focuses on our business activities.

IYA'S Cafe realizes that in the future when competition enters the market, additional revenues must be allocated for promotion in order to maintain market share.

As a retail establishment, IYA'S Cafe employs people to handle sales transactions. Computer literacy is a requirement for IYA'S Cafe employees. If an employee does not possess basic computer skills when they are hired, they are trained by our full-time technician. Our full-time technician is also available for customers in need of assistance. IYA'S Cafe's commitment to friendly, helpful service is one of the key factors that distinguishes IYA'S Cafe from other Internet cafes.

Sales forecast data is presented in the chart and table below.

Sales: IYA'S Cafe is basing their projected coffee and espresso sales on the financial snapshot information provided to them, were estimated by calculating the total number of hours each terminal will be active each day and then generating a conservative estimate as to how many hours will be purchased by consumers.

Cost of Sales: The cost of goods sold for coffee-related products was determined by the "retail profit analysis" we obtained from the survey. The cost of bakery items is 20% of the selling price. The cost of Internet access is $660 per month, paid to Millennium Studios Computers for networking fees. The cost of email accounts is 25% of the selling price.

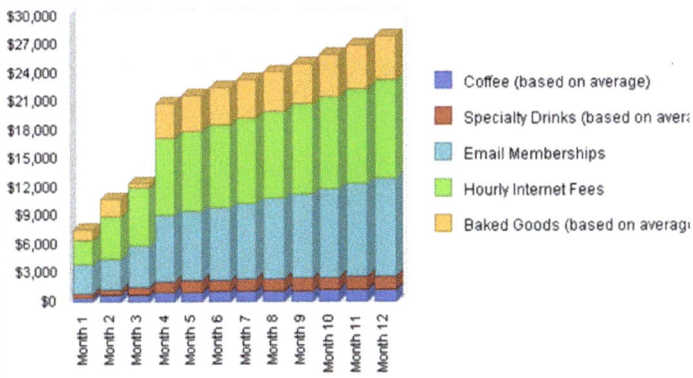

Sales Monthly

Legend:
- Coffee (based on average)
- Specialty Drinks (based on avera...)
- Email Memberships
- Hourly Internet Fees
- Baked Goods (based on averag...)

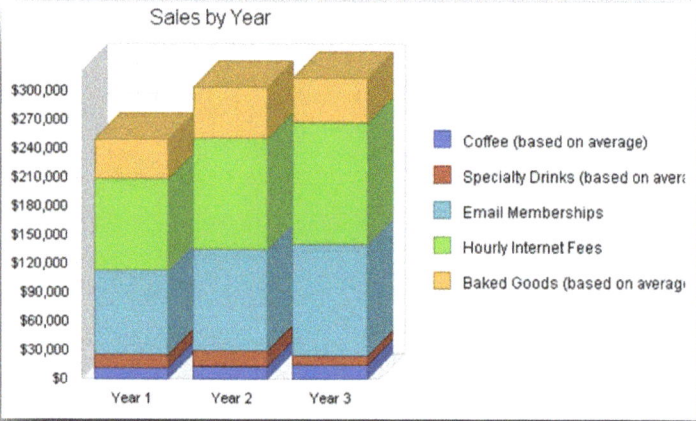

Sales by Year

$300,000
$270,000
$240,000
$210,000
$180,000
$150,000
$120,000
$90,000
$60,000
$30,000
$0

Year 1   Year 2   Year 3

- Coffee (based on average)
- Specialty Drinks (based on avera
- Email Memberships
- Hourly Internet Fees
- Baked Goods (based on average

Sales Forecast

	Year 1	Year 2	Year 3
Unit Sales			
Coffee (based on average)	12,016	14,068	15,475
Specialty Drinks (based on average)	6,654	7,913	8,705
Email Memberships	8,703	10,505	11,556
Hourly Internet Fees	38,269	46,365	51,002
Baked Goods (based on average)	32,673	46,365	51,002
Total Units Sales	98,315	121,001	133,103

Unit Prices

	Year 1	Year 2	Year 3
Coffee (based on average)	$1.00	$1.00	$1.00
Specialty Drinks (based on average)	$2.00	$2.00	$1.00
Email Membership	$10.00	$10.00	$10.00
Hourly Internet Fees	$2.50	$2.50	$2.50
Baked Goods (based on average)	$1.25	$1.25	$1.00

Sales

	Year 1	Year 2	Year 3
Coffee (based on average)	$12,016	$14,068	$15,5475
Specialty Drinks (based on average)	$13,308	$15,826	$8,705
Email Memberships	$87,030	$105,050	$115,560
Hourly Internet Fees	$95,673	$115,913	$127,505
Baked Goods (based on average)	$40,841	$52,688	$46,365
Total Sales	$248,868	$303,544	$313,610

Direct Unit Costs

	Year 1	Year 2	Year 3
Coffee (based on average)	$0.25	$0.25	$0.25
Specialty Drinks (based on average)	$0.50	$0.50	$0.25
Email Memberships	$2.50	$2.50	$2.50
Hourly Internet Fees	$0.63	$0.63	$0.63
Baked Goods (based on average)	$0.31	$0.31	$0.25

Direct Cost Sales

Coffee (based on average)	$3,004	$3,517	$3,869
Specialty Drinks (based on average)	$3,327	$3,957	$2,176
Email Memberships	$21,758	$26,263	$28,890
Hourly Internet Fees	$23,918	$28,978	$31,876
Baked Goods (based on average)	$10,210	$13,172	$11,591
Subtotal Direct Cost of Sales	$62,217	$75,886	78,403

The IYA'S Cafe management team has established some basic milestones to keep the business plan priorities in place. Responsibility for implementation falls on the shoulders of the owners. This milestones table below will be updated as the year progresses using the actual tables. New milestones will be added as the first year of operations commences.

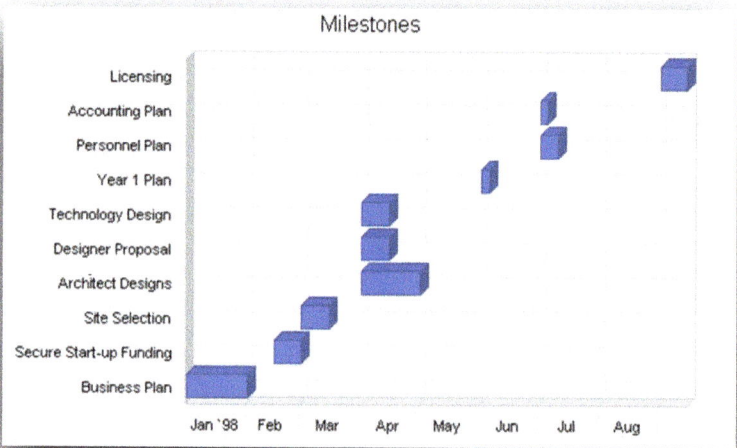

Milestones

Milestones

	Start Date	End Date	Budget	Manager
Business Plan	1/1/1998	2/1/1998	$1,000	Admin
Secure Start-up Funding	2/15/1998	3/1/1998	$1,000	Admin
Site Selection	3/1/1998	3/15/1998	$1,000	Admin
Architect Designs	4/1/1998	5/1/1998	$1,000	Admin
Designer Proposal	4/1/1998	4/15/1998	$1,000	Admin
Technology Design	4/1/1998	4/15/1998	$1,000	Admin
Year 1 Plan	6/5/1998	6/5/1998	$1,000	Admin
Personnel Plan	7/1/1998	7/10/1998	$1,000	Admin
Accounting Plan	7/1/1998	7/5/1998	$1,000	Admin
Licensing	9/1/1998	9/1/1998	$1,000	Admin
Total			$10,000	

Management Summary

The company, being small in nature, requires a simple organizational structure. Implementation of this organizational form calls for the owners to make all of the major management decisions in addition to monitoring all other business activities.

The staff will consist of six part-time employees working thirty hours a week at $5.50 per hour. In addition, TaNeHeSi (who is more technologically oriented to handle minor terminal repairs/inquiries) will deal with all technical concerns. This simple structure provides a great deal of flexibility and allows communication to disperse quickly and directly. Because of these characteristics, there are few coordination problems seen at IYA'S Cafe that are common within larger organizational chains. This strategy will enable IYA'S Cafe to react quickly to changes in the market.

Personnel Plan

	Year 1	Year 2	Year 3
Owner	$24,000	$26,400	$29,040
Part Time 1	$7,920	$7,920	$7,920
Part Time 2	$7,920	$7,920	$7,920
Part Time 3	$7,920	$7,920	$7,920
Part Time 4	$7,920	$7,920	$7,920
Part Time 5	$7,920	$7,920	$7,920
Part Time 6	$3,960	$7,920	$7,920
Technician	$21,731	$23,904	$26,294
Manager	$4,000	$24,000	$26,400
Total People	9	9	9
Total Payroll	$93,291	$121,824	$129,254

Financial Plan

The following sections lay out the details of our financial plan for the next three years.

This business plan is prepared to obtain financing in the amount of $24,000. The supplemental financing is required to begin work on-site preparation and modifications, equipment purchases, and to cover expenses in the first year of operations.

Additional financing has already been secured as follows:

1. $24,000 from a business loan

2. $19,000 from personal savings

3. $36,000 from three investors

4. $9,290 in the form of short-term loans

Start-up Funding

Start-up Expenses to Fund $62,290

Start-up Assets to Fund $26,000

Total Funding Required $88,290

Assets

Non-cash Assets from Start-up $2,000

Cash Requirements from Start-up $24,000

Additional Cash Raised $0

Cash Balance on Starting Date $24,000

Total Assets $26,000

Liabilities and Capital

Liabilities

Current Borrowing $9,290

Long-term Liabilities $24,000

Accounts Payable (Outstanding Bills) $0

Other Current Liabilities (interest-free) $0

Total Liabilities $33,290

Capital

Planned Investment

Cale Bruckner $19,000

Luke Walsh $12,000

Doug Wilson $12,000

John Underwood $12,000

Additional Investment Requirement $0

Total Planned Investment $55,000

Loss at Start-up (Start-up Expenses) ($62,290)

Total Capital ($7,290)

Total Capital and Liabilities $26,000

Total Funding $88,290

7.2 Important Assumptions

Basic assumptions are presented in the table below.

General Assumptions

	Year 1	Year 2	Year 3
Plan Month	1	2	3
Current Interest Rate	8.00%	8.00%	8.00%
Long-term Interest Rate	10.00%	10.00%	10.00%
Tax Rate	30.00%	30.00%	30.00%
Other	0	0	0

7.3 Key Financial Indicators Important benchmark data is presented in the chart below.

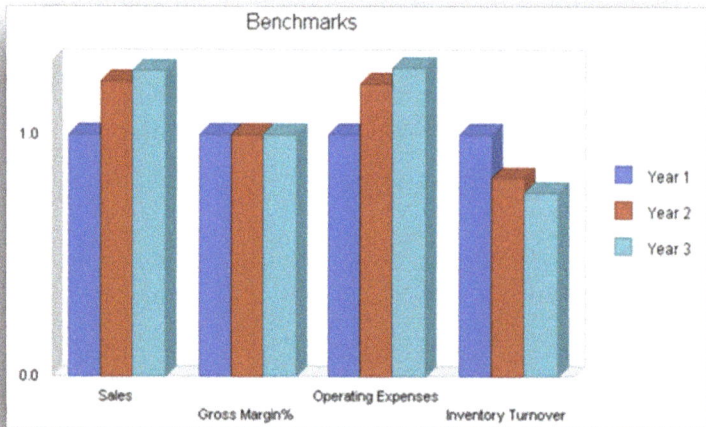

Break-even data is presented in the chart and table below.

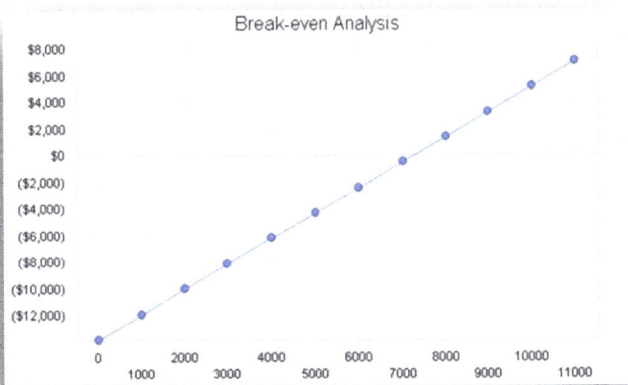

Break-even Analysis

Monthly Units Break-even $7,294

Monthly Revenue Break-even $18,462

Assumptions:

Average Per-Unit Revenue $2.53

Average Per-Unit Variable Cost $0.63

Estimated Monthly Fixed Cost $13,847

Payroll Expense: The founders of IYA'S Cafe, Cale Bruckner, will receive salaries based on sales. IYA'S Cafe intends to hire five or six part-time employees by the end of year one and a full-time technician at standard AfRakAn hourly rates.

Rent Expense: IYA'S Cafe is leasing a 1700 square foot facility at $.85/sq. foot. The lease agreement IYA'S Cafe signed specifies that we pay a to be determined amount for a total of 36 months. At the end of the third year, the lease is open for negotiations, and IYA'S Cafe may or may not re-sign the lease depending on the demands of the lessor.

Utilities Expense: As stated in the contract, the lessor is responsible for the payment of utilities, including gas, garbage disposal, and real estate taxes. The only utilities expense that IYA'S Cafe must pay is the phone bill generated by fifteen phone lines; thirteen will be dedicated to modems and two for business purposes. The basic monthly service charge for each line provided by US West is $17.29. The 13 lines used to connect the modems will make local calls to the network provided by Millennium Studios resulting in a monthly charge of $224.77. The two additional lines used for business communication will cost $34.58/month plus long-distance fees. IYA'S

Cafe assumes that it will not make more than $40.00/month in long-distance calls. Therefore, the total cost associated with the two business lines is estimated at $74.58/month, and the total phone expense at $299.35/month.

Marketing Expense: IYA'S Cafe will allocate $33,750 for promotional expenses over the first year. These dollars will be used to advertise in local newspapers to build consumer awareness. For additional information, please refer to section 5.0 of the business plan.

Insurance Expense: IYA'S Cafe has allocated $1,440 for insurance for the first year. As revenue increases in the second and third year of business, IYA'S Cafe intends to invest more money for additional insurance coverage.

Depreciation: In depreciating our capital equipment, IYA'S Cafe used the Modified Accelerated Cost Recovery Method. We depreciated our computers over a five-year time period and our fixtures over seven years.

Taxes: IYA'S Cafe is an LLC, and, as an entity, it is not taxed. However, there is a 15% payroll burden.

Chapter 7
CONCLUSION

In closing, while many other elements probably deserve attention in this regard, our goal has been to provide a detailed look at the dynamic aspects involved in building a CyberCafe in AfRaKa that's based on sound financial practices and the need to take advantage of the strong possibilities for success in a country like Rwanda. A fundamental principle to consider when investigating in any society is that it shouldn't be driven by exploitation but by the principles of harmony, which will also promote the well-being of the community that one chooses to establish a business in. In this regard, the investor needs to consider themselves as a member of the society where the company is located and aims to provide quality services for that community.

The ability to provide quality and reliable services can lead to franchise possibilities not only in Rwanda but also in other parts of Africa. So as Iya Café grows, it should take advantage of opportunities to expand and prosper while serving as a new-age Business model that is driven by the principles of Kimoyo as defined in the books by my won arakurunrin @Kofi PieSie and @Ini-Herit Shawn Kalfani and their Research Teams entitled Beautiful Lessons about Kimoyo and Spear Masters: A Historical Survey of the Minds of African Warrior Scholars Vol. 4. The gradual return to the traditions and customs of our Ancestors will be difficult but not impossible. Still, we must refuse attempts to impose colonial mindsets on the indigenous peoples of AfRaKa.

For this to be achieved, we must encourage young minds to become familiar with scientific methods and the studies of History, Political Science, Medicine, Finance, and other fields. This will empower a generation to move forward and address the immediate concerns of their won Ebi and their communities.

BIBLIOGRAPHY

•AfRaKan Academy of Sciences. Workshop on Science and Technology Communication Networks in AfRaKa. Nairobi: AfRaKan Academy of Science, 1993

•Balaam, David N., and Michael Veseth, eds. Introduction to International Political Econo New Jersey: Prentice Hall, 1996

•Baradat, Leon P. Political Ideologies. New Jersey: Prentice Hall, 1994

•Billet, Bret L. Investment behavior of Multinational Corporations in Developing Areas. New Brunswick: Transaction Publishers, 1991

•Clough, Michael. Free at Last?. New York: Council on Foreign Relations Press, 1992

•Drew, Eileen P., and F. Gordon Foster, eds. Information Technology in Selected Countries. Tokyo: United Nations University, 1994

•Dubois, W.E.B. The World and AfRaKa. New York: International Publishers, 1965

•Fieldhouse, D.K.. Black AfRaKa: 1940 1980. Boston: Unwin Hyman, 1986

•Haggard, Stephan, and Robert R. Kaufman, eds. The Politics of Economic Adjustment. Princeton: Princeton University Press, 1992

•Harbeson, John W., and Donald Rothchild, eds. Afiica in World Politics. Boulder: Westview Press, 1991

•Leedy, Paul D. Practical Research: Planning and Design . 5th ed. New York: Macmillan Publishing Company, 1993

•Moran, Theodore H. Multinational Corporations. Massachusetts: Lexington Books, 1985

•National Research Council, Office of International Affairs, Bridge Builders. Washington: National Academy Press, 1996

•Piesie, Kofie., Beautiful Lessons About Kimoyo, Same Tree Different Branch Publishing, U.S.A 2021

•Rodney,Walter. How Europe Underdeveloped AfRaKa. Washington: Howard University Press, 1974

•Rosenbloom, Richard. Technology and Information Transfer. Boston: Havard University Press, 1970

•Sandbrook, Richard. The Politics of AfRaKa's Economic Recovery. New York: Cambridge University Press, 1993

•Shibre, Zewdie, and Abdulhamid Bedri, eds. Regional Development Problems in AfRakA . Addis

•Ababa: Institute of Development Research, 1993

•Slater, Robert O., Barry M. Schutz, and Steven R. Dorr, eds., Global Transformation and the Third World. Boulder: Lynne Rienner Publishers, 1992

•Turabian, Kate. A Manual for Writers. 5th ed. Chicago: University of Chicago, 1987

•Weiss, Thomas G., and Merl A. Kessler, eds. Third World Security in the Post Cold War Era Boulder: Lynne Reinner Publishers, 1991

•Weston, Alan F. Information Technology in a Democra . Cambridge: Harvard University Press,1971

•Articles, Papers and Public Documents da Costa, Peter. "AfRakA Communication: Internet A Statist Model", Addis Ababa: International Press Service, 10 September 1996,

•National Telecommunications and Information Administration, "U.S. Goals and Objectives for the Information Society and Development Conference", prepared remarks of Vice President Al Gore, delivered via satellite to the Information Society and Development Conference in Midrand, South AfRaKa(May 13, 1996)

•Semret, Nemo. "Unleashing AfRakA's Potential: The Technological Reasons for Open and Competitive Cybercommunications", a paper delivered at The Second Annual Meeting of the AfRakA Scientific Society, Washington, June 22, 1996

•Burka, Lauren P. "A Hypertext History of Multi-User Dimensions." MUD History 1993. http://www.utopia.com/talent/lpb/muddex/essay (2 Aug. 1996).

•Fine Arts." Dictionary of Cultural Literacy. 2nd ed. Ed. E. D. Hirsch, Jr., Joseph F. Kett, and James Trefil. Boston: Houghton Mifflin. 1993. INSO Corp. America Online. Reference Desk/Dictionaries/Dictionary of Cultural Literacy (20 May 1996).

•Senet: https://senet.cloud/en/blog/how-to-open-cyber-cafe

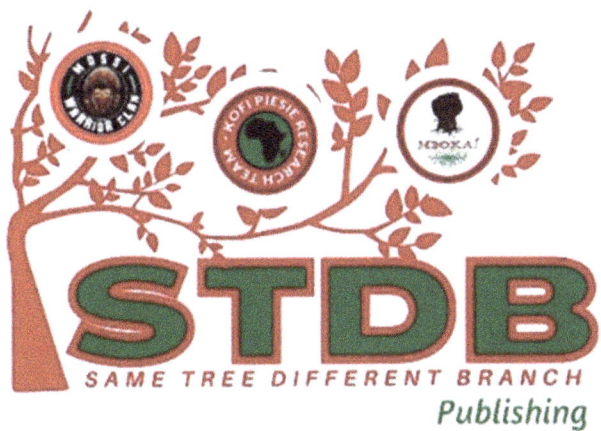

www.ingramcontent.com/pod-product-compliance
Lightning Source LLC
Chambersburg PA
CBHW042118190326
41519CB00030B/7537